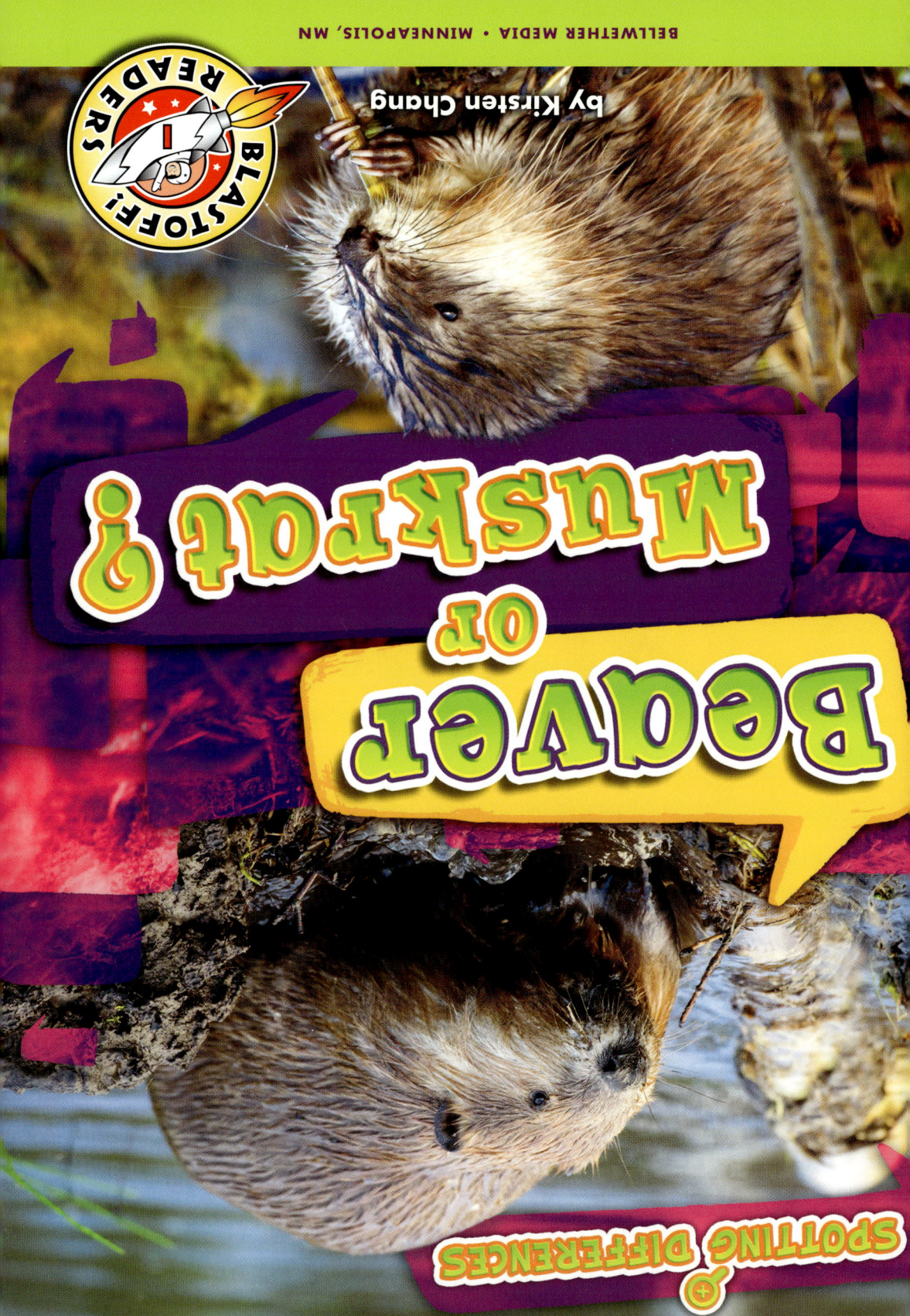

Beaver or Muskrat?

SPOTTING DIFFERENCES

by Kirsten Chang

BELLWETHER MEDIA • MINNEAPOLIS, MN

This edition first published in 2021 by Bellwether Media, Inc.

No part of this publication may be reproduced in whole or in part without written permission of the publisher. For information regarding permission, write to Bellwether Media, Inc., Attention: Permissions Department, 6012 Blue Circle Drive, Minnetonka, MN 55343.

Library of Congress Cataloging-in-Publication Data

Names: Chang, Kirsten, 1991- author.
Title: Beaver or muskrat? / by Kirsten Chang.
Description: Minneapolis, MN : Bellwether Media, 2021. | Series: Blastoff! readers: spotting differences | Includes bibliographical references and index. | Audience: Ages 5-8 | Audience: Grades K-1
Summary: "Developed by literacy experts for students in kindergarten through grade three, this book introduces beavers and muskrats to young readers through leveled text and related photos"-- Provided by publisher.
Identifiers: LCCN 2019053748 (print) | LCCN 2019053749 (ebook) | ISBN 9781644871973 (library binding) | ISBN 9781681038216 (paperback) | ISBN 9781618919557 (ebook)
Subjects: LCSH: Beavers--Juvenile literature. | Muskrat--Juvenile literature.
Classification: LCC QL737.R632 C45 2021 (print) | LCC QL737.R632 (ebook) | DDC 599.37--dc23
LC record available at https://lccn.loc.gov/2019053748
LC ebook record available at https://lccn.loc.gov/2019053749

Text copyright © 2021 by Bellwether Media, Inc. BLASTOFF! READERS and associated logos are trademarks and/or registered trademarks of Bellwether Media, Inc.

Editor: Elizabeth Neuenfeldt   Designer: Jeffrey Kollock

Printed in the United States of America, North Mankato, MN.

★ **Blastoff! Universe**

Reading Level

Grade K — Blastoff! Beginners
Grades 1-3 — Blastoff! Readers
Grade 4 — Blastoff! Discovery

LEVELS

Level 1 provides the most support through repetition of high-frequency words, light text, predictable sentence patterns, and strong visual support.

Level 2 offers early readers a bit more challenge through varied sentences, increased text load, and text-supportive special features.

Level 3 advances early-fluent readers toward fluency through increased text load, less reliance on photos, advancing concepts, longer sentences, and more complex special features.

Blastoff! Readers are carefully developed by literacy experts to build reading stamina and move students toward fluency by combining standards-based content with developmentally appropriate text.

Table of Contents

Beavers and Muskrats	4
Different Looks	8
Different Lives	12
Side by Side	20
Glossary	22
To Learn More	23
Index	24

Beavers and Muskrats

Beavers and muskrats are **mammals**. They swim and build their homes in water.

beavers

9

Both animals have brown fur and long tails. What makes them different?

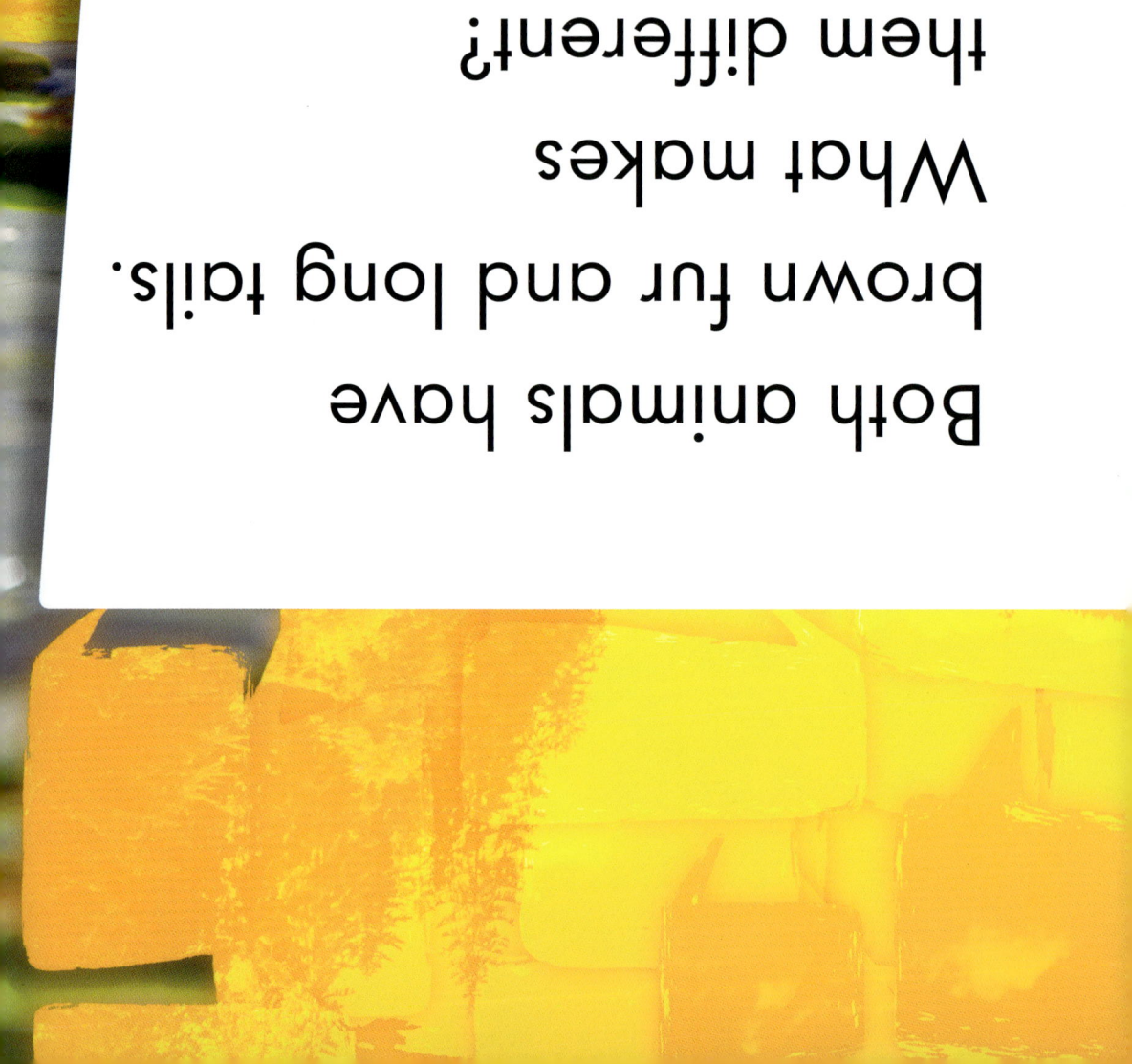

muskrat

Different Looks

Muskrats are small. A beaver can weigh 20 times more than a muskrat!

6

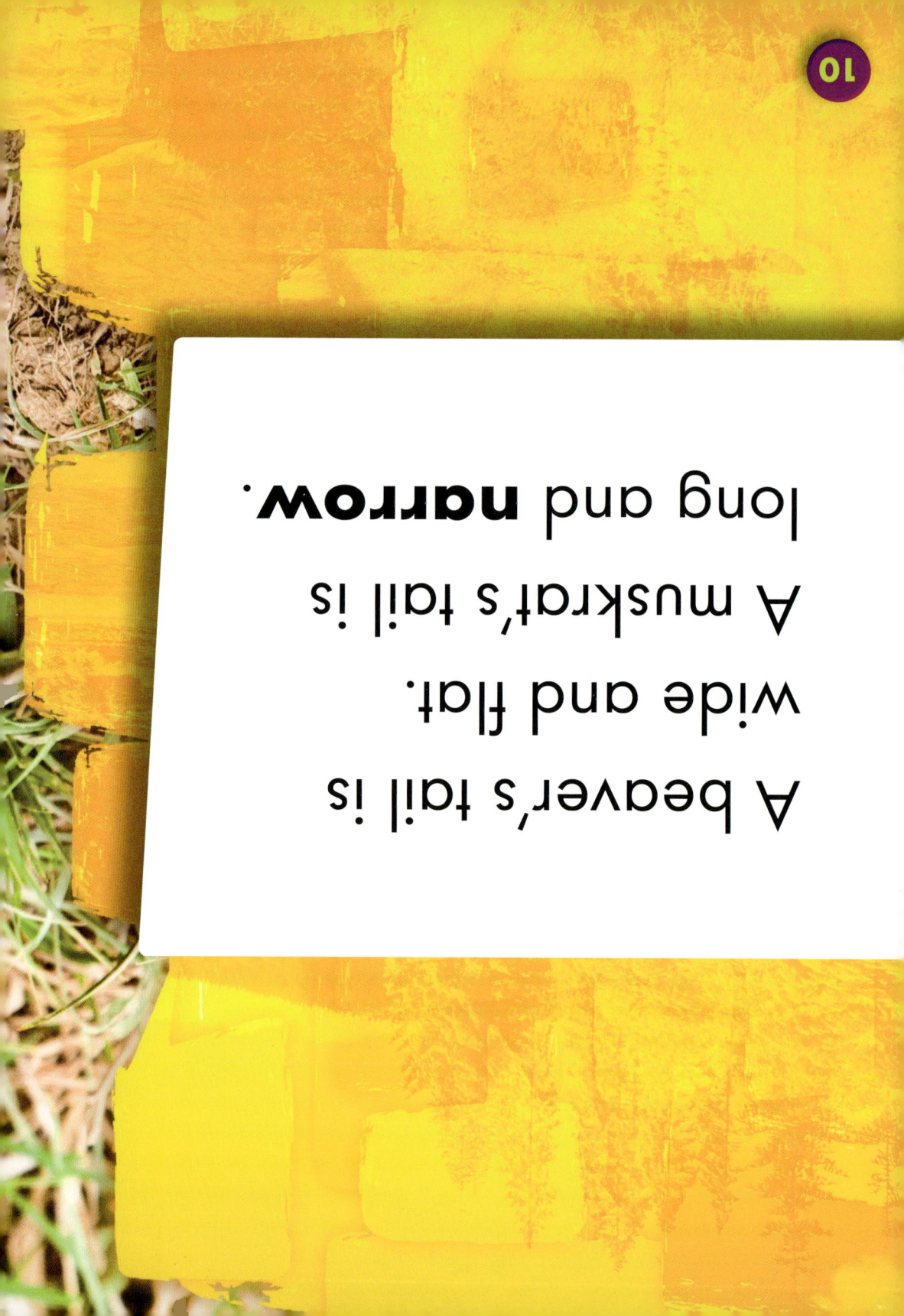

A beaver's tail is **wide and flat**.
A muskrat's tail is **long and narrow**.

narrow tail

11

Different Lives

Beavers build **dams** that make ponds. Then they build **lodges** for homes.

Muskrats do not build dams. Their lodges are smaller than beaver lodges.

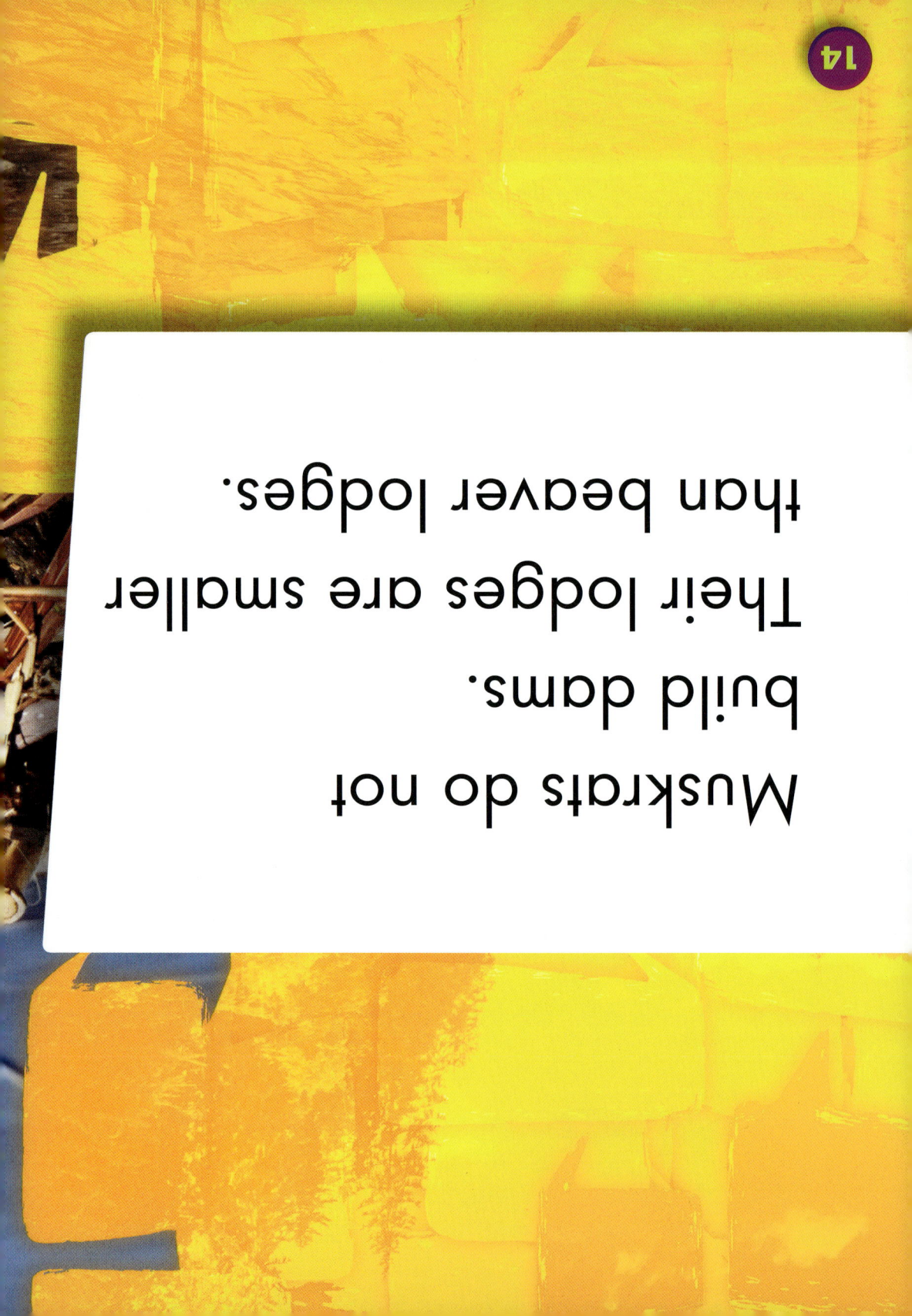

muskrat lodge

15

Muskrats eat plants and small animals such as **crayfish**. Beavers eat tree bark and wood.

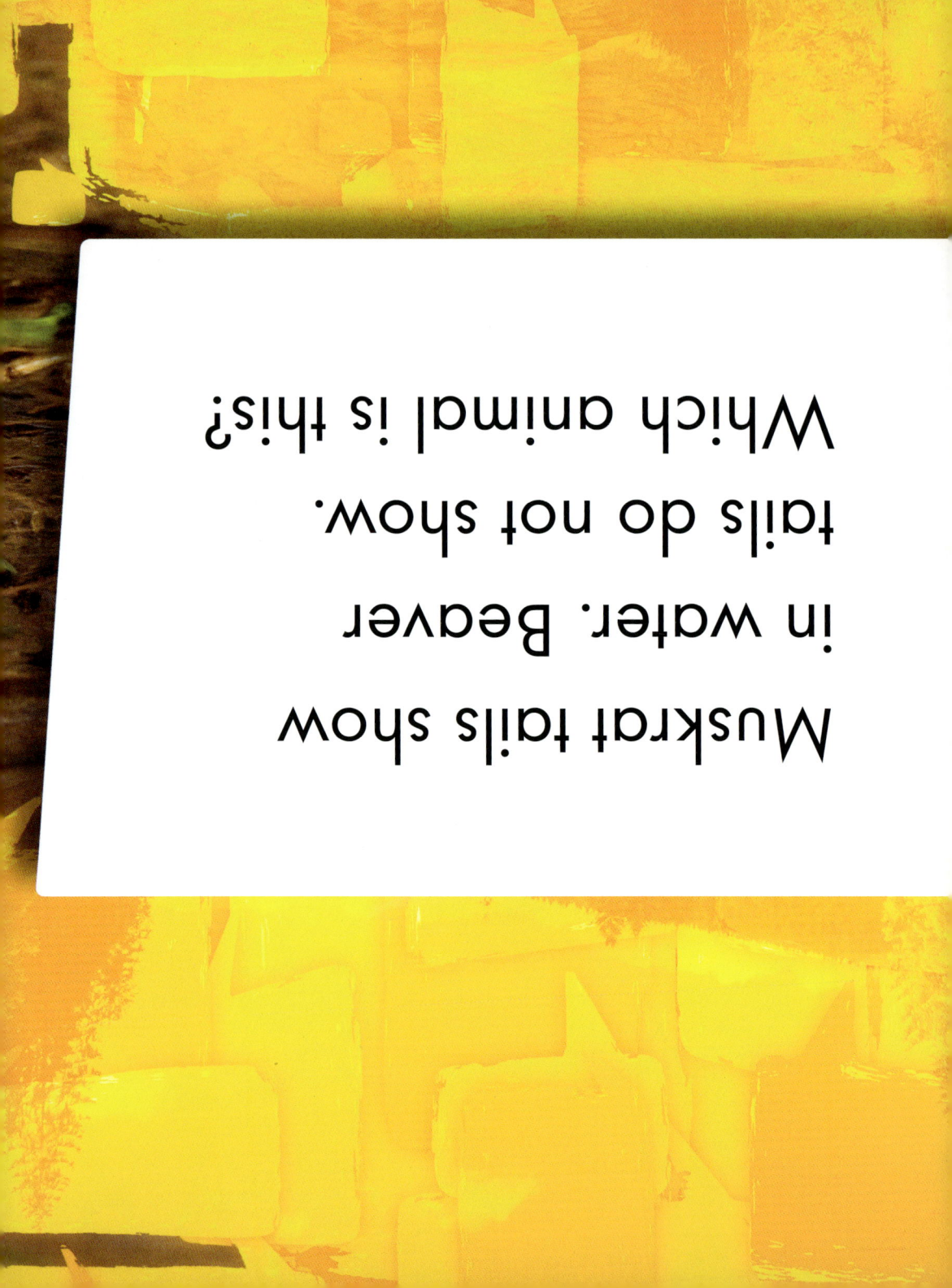

Muskrat tails show in water. Beaver tails do not show. Which animal is this?

Side by Side

large, round body

wide, flat tail

Beaver Differences

- tails do not show in water
- build dams and lodges
- eat tree bark and wood

20

small, round body

long, narrow tail

Muskrat Differences

eat plants and small animals

build small lodges to live in

tails show in water

Glossary

crayfish

small animals that look like lobsters and live in rivers

dams

walls built across rivers or streams that stop water flow

lodges
homes for beavers or muskrats

mammals

warm-blooded animals that have hair and feed their young milk

narrow

thin

To Learn More

AT THE LIBRARY

Gaertner, Meg. *Muskrats*. Minneapolis, Minn.: Pop!, 2019.

Grack, Rachel. *Beavers*. Minneapolis, Minn.: Bellwether Media, 2020.

Kissock, Heather. *Beavers*. New York, N.Y.: AV2 by Weigl, 2019.

ON THE WEB

FACTSURFER

Factsurfer.com gives you a safe, fun way to find more information.

1. Go to www.factsurfer.com.

2. Enter "beaver or muskrat" into the search box and click 🔍.

3. Select your book cover to see a list of related content.

Index

animals, 6, 16, 18

bark, 16

build, 4, 12, 14

crayfish, 16

dams, 12, 13, 14

fur, 6

homes, 4, 12

lodges, 12, 13, 14, 15

mammals, 4

plants, 16

ponds, 12

swim, 4

tails, 6, 10, 11, 18

water, 4, 18

weigh, 8

wood, 16

The images in this book are reproduced through the courtesy of: Ghost Bear, front cover; Jukka Jantunen, front cover, p. 11 (bubble); BMJ, pp. 4-5, 16-17; Toni Genes, pp. 6-7; Robert McGouey/Wildlife/ Alamy, pp. 8-9, 22 (mammals); blickwinkel/ Alamy, pp. 10-11, 18-19; O Brasil que poucos conhecem, pp. 12-13; Enrique Aguirre, p. 13 (lodge); Mike Truchon, pp. 14-15; jmarino, pp. 15 (lodge), 21 (middle); Jody Ann, p. 20 (beaver); Danita Delmont, p. 20 (left); Filip Fuxa, p. 20 (middle); Tom Worsley, p. 20 (right); Sergey Bond, p. 21 (muskrat); Sergey Uryadnikov, p. 21 (left); Vishnevskiy Vasily, p. 21 (right); Miroslav Hlavko, p. 22 (crayfish); Nikki_N, p. 22 (dams); Marilyn D. Lambertz, p. 22 (lodges); Merrimon Crawford, p. 22 (narrow).